Contents

Stem Cells and Cloning

Two of the most controversial topics in the modern history of biology and medicine are stem cell research and cloning. They are both exciting new areas of research, holding possible promises of cures for a wide range of diseases, but also thorny ethical issues that make us confront some of the most basic questions in life. What does it mean to be human? When do we become human, or a person? What trade-offs are we willing to make in our quest to cure disease and enable healthier lives for those who suffer? Are there alternative research paths, and should science be allowed to follow any path or should there be limits on scientific investigation? These are just a few of the questions raised by stem cell and cloning research. We're now entering the Biological Revolution, where we may be able to provide not only miraculous medical cures, but also perhaps redesign ourselves in ways we never thought possible before. The decisions made now can affect the entire future of humanity. But first we need to know more about stem cells and about cloning. What are the scientific facts, and how do these facts affect the decisions and outcomes? Welcome to Stem Cells and Cloning!

WHY IS EVERYONE SO INTERESTED IN STEM CELLS?

Most people hadn't heard about stem cells until 1998, when James Thomson of the University of Wisconsin–Madison first isolated human embryonic **stem cells**. Suddenly *everyone* was interested in stem cells, and they were a hot topic not only for scientists but also for politicians, patient advocates, and many others. The reason is that stem cells may hold the key to treatment of many of the major diseases we currently face, diseases that kill and disable millions of people. In the past, infectious diseases were the main scourge of mankind—polio, plague, smallpox, and so on. Better hygiene and the discovery and use of antibiotics and vaccines have greatly decreased the problems caused by such diseases. But control over most infectious diseases has now led to other diseases being the chief health problems in industrialized nations—degenerative diseases.

Degenerative diseases involve the slow, gradual breakdown of tissues and organs, or are the result of death of some of the cells in an organ, which

Rank (1998)	Cause of Death	Number of Deaths	Death Rate per 100,000	Percent of Total Deaths
1	Heart disease	724,859	268.2	31.0
2	Cancer	541,532	200.3	23.2
3	Stroke	158,448	58.6	6.8
4	Chronic obstructive lung disease	112,584	41.7	4.8
5	Accidents	97,835	36.2	4.2
6	Pneumonia/influenza	91,871	34.0	3.9
7	Diabetes	64,751	24.0	2.8
8	Suicide	30,575	11.3	1.3
9	Kidney disease	26,182	9.7	1.1
10	Chronic liver disease and cirrhosis	25,192	9.3	1.1
11	Septicemia		8.8	1.0
12	Alzheimer's disease		8.4	1.0
13	Homicide and legal intervention		6.8	0.8
14	Atherosclerosis		5.7	0.7
15	Hypertension		5.3	0.6

weakens the organ and leads to further breakdown and eventually organ failure. These are the major killers in the U.S. and all industrialized nations. Heart disease, cancer, stroke, chronic lung and liver disease, diabetes, Alzheimer's disease are all in the list (See Table 1). For example, when someone has a heart attack, the entire heart doesn't die; instead, *part* of the heart muscle dies due to lack of oxygen (from a blocked or obstructed blood vessel.) The heart muscle is weakened and can lead to further problems and death. The hope is that stem cells could be used to repair and replace the damaged heart muscle, essentially "rejuvenating" the heart and restoring it to full function. The same idea for replacing damaged or dead tissues could be used for other diseases, including some neurological diseases not in the top 15 list, such as Parkinson's disease, Huntington's disease, ALS (Lou Gehrig's disease), and spinal cord injuries.

Stem cells would not be used to grow whole organs in culture for transplant.

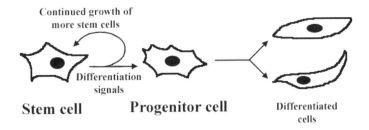

Continued growth of
more stem cells

Differentiation
signals

Stem cell   **Progenitor cell**   Differentiated
cells

FIGURE 1. Characteristics of a stem cell.

First, it's just not possible or practical to do that. The more direct route, of injecting the stem cells and having them repair and replace only the damaged part of an organ, is not only more effective but also easier.

WHAT'S A STEM CELL?

What makes a stem cell such an attractive candidate for repairing damaged organs and tissues? A stem cell has two basic characteristics: (1) it continues to grow and proliferate, maintaining a pool of cells for possible use, and (2) given the correct signals (for example, stimulation by a hormone or "growth factor" such as insulin, steroid hormones, or nerve growth factor), it can differentiate into a particular specialized cell type (Figure 1). It often goes through various steps to do this, first forming a **progenitor cell**, a precursor to the specific cell type, then changing into a **differentiated cell** (a specialized cell type that carries out a specific function in the body, such as a heart muscle cell, a neuron in the brain, or a red blood cell to carry oxygen to other cells in the body.)

Some stem cells seem to have more abilities, or possibilities, than other stem cells, termed the **potency** of the stem cell. A stem cell that is **unipotent** can form only one differentiated cell type. A **multipotent** stem cell can form multiple different cell and tissue types. A **pluripotent** stem cell can form most or all of the 210 differentiated cell types in the adult body. A **totipotent** stem cell can form not only all adult body cell types, but also the specialized tissues needed for development of the embryo, such as the placenta.

WHERE DO WE GET STEM CELLS?

There are actually lots of sources for stem cells (Figure 2). In the past it was thought that only a few adult tissues contained stem cells, and that they were very limited in the types of cells that they could form. For example, it has been known for over 20 years that bone marrow contains stem cells that make all of the blood cells (red cells to carry oxygen, white cells for immunity, etc.), termed **hematopoietic** (blood forming) cells, and that the intestine contains stem cells that continually regenerate the intestinal lining. Other well-known examples include skin stem cells and stem cells that make sperm.

Another type of stem cell that has been known about for many years is the **embryonal carcinoma cell (EC cell)**. This cell actually comes from a tumor called a **teratoma** (if benign) or **teratocarcinoma** (if malignant), which forms when a germ cell such as an oocyte (an egg) spontaneously starts to grow and divide. Scientists noticed that occasionally such tumors contained not just a disorganized mass of growing cells, as in most tumors, but also some differentiated tissues, such as a bit of bone, hair, or teeth! This led to research in which the embryonal carcinoma cells were grown over a period of years and "tamed," so that they would not grow as disorganized tumor masses but instead form specific differentiated cell types such as nerves. This work also led to the recognition that the early embryo contains similar types of stem cells.

The **embryonic stem cell (ES cell)** has been called the master cell of the early embryo. This is one of the first cells formed during early development of the embryo, and these are the cells that go on to form all of the tissues of the body. Embryonic stem cells were first isolated from mouse embryos in 1981, and from human embryos in 1998 by Dr. James Thomson and his colleagues at the University of Wisconsin in Madison. Scientists found that, as with embryonal carcinoma cells, they could sometimes get the embryonic stem cells to form different specialized tissues in the laboratory culture dish. In humans, embryonic stem cells form about 5 to 7 days after conception. At that point in development, the human embryo, or a **blastocyst**, looks like a hollow ball with some cells inside. The outer layer of cells are the **trophoblast** cells, which go on to form the placenta. The cells inside are called the **inner cell mass**—these are the embryonic stem cells (Figure 2).

At almost the same time in 1998 as the announcement that human embryonic stem cells were first isolated, Dr. John Gearhart and his team at Johns Hopkins University reported that they had isolated what they called human **embryonic germ cells (EG cells)**. These cells are derived from the progenitor cells that will become germ cells (egg or sperm) from early fetal tissue (5 to 9 weeks after conception.) When placed into laboratory culture dishes, the cells showed many of the same characteristics as embryonic stem cells. That is, they seem to be pluripotent, able to form most or all of the tissues of the adult body.

As human development proceeds and various tissues start to form, the embryonic stem cells form progenitor cells, the partially specialized cells that go on to form the specific differentiated tissues of the body. Since this is a gradual process, it's not surprising to find that tissues in the developing fetus also contain stem cells that can form several different cell types.

What has been surprising is the recognition in recent years that even fully

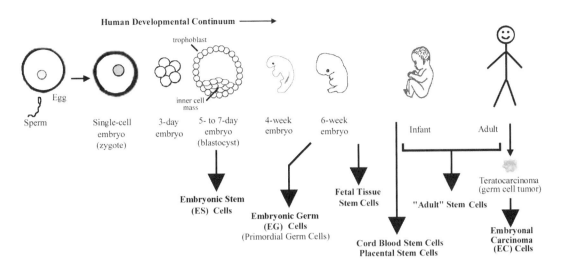

FIGURE 2. Stem cell sources.

formed adult tissues also contain stem cells. The surprise has been that not only do the few tissues mentioned before (such as bone marrow) contain stem cells, but most or all adult tissues contain stem cells. A further surprise was finding that at least some of these **adult stem cells** can form many more different types of tissue than was previously thought. The term adult stem cell is actually not completely correct, because these cells are present in organs and tissues from the moment we are born. In fact, it's now recognized that similar stem cells are found in the **placenta** and in **umbilical cord blood**. Since all of these stem cells are found in various body tissues at or after birth, some people term them **tissue stem cells** or **non-embryonic stem cells**. However, some people lump all of these tissue stem cells together under the term adult stem cells, to distinguish them from embryonic stem cells.

Most of the attention in terms of possible use for treatment of degenerative diseases has focused on embryonic stem cells and the group collectively termed adult stem cells. What are the advantages and disadvantages of these stem cell sources to treat disease, and why is there such controversy over stem cells? Let's examine embryonic and adult stem cells more closely.

EMBRYONIC STEM CELLS

Embryonic stem cells are **pluripotent**, meaning that they can form all of the tissues of the adult body. This is what they do during normal embryological

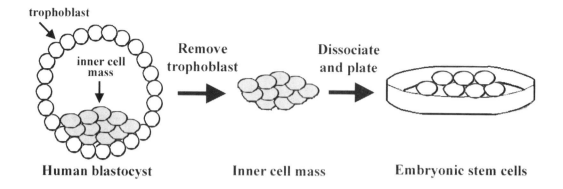

FIGURE 3. Isolation of embryonic stem cells.

development; their "job description" is the initial formation of all the body tissues. As mentioned before, mouse embryonic stem cells were first isolated in 1981, while human embryonic stem cells were first isolated in 1998. Embryonic stem cells are isolated by breaking open an embryo and removing the inner cells (Figure 3). This process necessarily destroys the embryo, and this is the main reason embryonic stem cells are so controversial—to isolate human embryonic stem cells, human embryos are destroyed. But if embryonic stem cells can perform all the wonders claimed for them in tissue regeneration, might it be acceptable that some human embryos are destroyed so that millions of lives can be spared? Let's first examine the scientific facts, then we'll come back to the ethical question.

One possible source for human embryonic stem cells is excess embryos from **assisted reproductive technologies**, such as **in vitro fertilization (IVF)**. Infertile couples have their eggs and sperm mixed in a laboratory dish, producing many embryos. Some embryos are implanted into the woman's womb to initiate a pregnancy, while the rest are frozen for future use. In some cases the couple never implant all of the embryos, and these "leftover" embryos may be donated (with the couple's consent) for research.

Human embryonic stem cells have been touted as a "virtual fountain of youth," because of their potential to repair and rejuvenate any damaged tissue in the body. In theory this should be possible, because these are the cells that initially form all body tissues during embryonic development. Remember the two characteristics of stem cells—continued growth and the ability to form differentiated tissues when given the correct signal. The advantages of embryonic stem cells are supposedly that they can be grown indefinitely in laboratory cul-

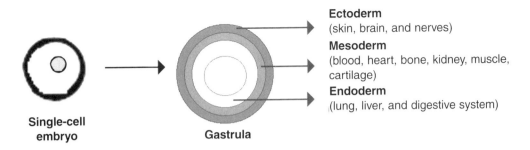

Ectoderm
(skin, brain, and nerves)
Mesoderm
(blood, heart, bone, kidney, muscle, cartilage)
Endoderm
(lung, liver, and digestive system)

Single-cell embryo

Gastrula

FIGURE 4. The three primary germ layers during embryo development.

tures, providing an almost limitless source of stem cells, and that they can form any body tissue. The published scientific papers on human embryonic stem cells back up the first claim—apparently limitless growth in the lab dish.

Tests for Pluripotency, the Ability to Form Any Body Tissue

What's the basis for the claim that an embryonic stem cell can form any adult tissue? It's based on several different types of scientific studies. First, the simple fact that under normal developmental conditions when left alone to do their job, embryonic stem cells go ahead to form all the tissues of our bodies! Of course, that's assuming you leave the cells in the intact embryo . . .

Second, one test used to demonstrate that embryonic stem cells can form many different tissues is to inject them into mice that are **immunodeficient**, meaning that they lack a functional immune system. Such mice need to be kept in special sterile environments so that they do not get an infection because without a functional immune system they would not be able to fight off even a cold, and would die quickly. But their lack of an immune system means that they also will not reject transplanted tissues, so they are a good way to test the ability of transplanted cells to grow inside the body. When embryonic stem cells are placed into these immunodeficient mice, they form tumors. The tumors are similar to teratomas, with some of the embryonic stem cells differentiating into specialized cell types and tissues. The tumors are examined closely to see what types of specialized cells grow.

In particular, scientists look to see if cells grow that represent each of the **three primary germ layers** (Figure 4). During embryological development, the embryonic stem cells first form three semi-specialized layers, the primary germ layers. Cells from each layer then go on to form specific tissues in the body.

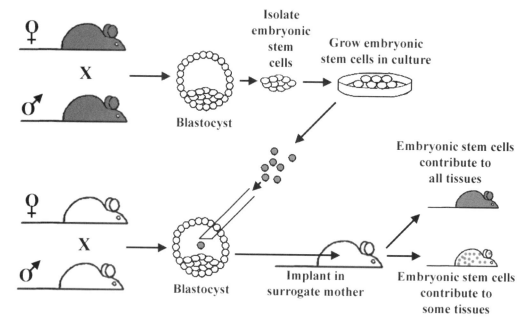

FIGURE 5. Evidence that an embryonic stem cell can form all tissues.

Ectoderm forms tissues like skin, brain, and nerves, **mesoderm** forms tissues like blood, heart, bone, kidney, muscle, and cartilage, and **endoderm** forms tissue such as lung, liver, and digestive system. If the injected embryonic stem cells form, for example, nerve cells, heart cells, and intestinal cells, this indicates they should be pluripotent, able to form almost all tissues, because the three cell types mentioned each come from a different primary germ layer. The fact that they do not form every single tissue is not so important as the indication that they have the ability to form a range of tissue types. Of course, this is not definite proof, but only an indication that the cells have this ability.

Third, in mouse studies embryonic stem cells have been grown in culture, then injected back into an early mouse embryo (blastocyst; Figure 5). If the embryonic stem cell was from a black mouse, and the cell was put into an embryo for a white mouse, then the coat color of the mouse once born (as well as other genetic markers) can be tracked to see what tissues the injected embryonic stem cell helped to form. In this situation, mouse embryonic stem cells have been able to help construct most or all of the body tissues of the mouse born. This experiment is considered the "gold standard" for testing the ability of any stem cell to form all tissues of the body. One problem to keep in mind with this experiment is that we are putting the stem cell back into a normal developmental environment, inside an embryo, where it was exposed to all of the normal signals that cells experience.

What Is the Hope for Embryonic Stem Cells?

The hope for embryonic stem cells is that large numbers of the cells can be grown and then given chemical and hormonal signals to specialize into a specific desired tissue. This might make it possible to use stem cells to create new cells and tissues on demand, to alleviate or even possibly cure a myriad of human diseases. In this way, specific nerve cells could be generated to implant into Parkinson's patients, or different nerve cells for Alzheimer's patients, or insulin-secreting pancreas cells for diabetics, or spinal cord neurons to repair spinal cord injury. Theoretically the possibilities are endless; depending on the specific cell needed, the proper signal would be given to the dish of cells, all the cells would form the desired differentiated cell type, and the cells would be transplanted into the appropriate patient to repair and regenerate the damaged tissue.

So far there is evidence that embryonic stem cells in the lab culture dish can be persuaded to form various cell types, including beating heart cells, muscle, blood, and different types of nerve cells. In studies in animals, there are indications that embryonic stem cells can form nerve cells, partially repair spinal cord injury, and provide modest relief from some of the symptoms of Parkinson's disease. However, the research is still many years away from any treatments for human patients.

What Are the Problems with Embryonic Stem Cells?

One scientific problem with embryonic stem cells is the inefficiency in getting them to start growing and keep growing in the laboratory. Only about 1 in 10 of the embryos used actually provide useable stem cells that will take hold in culture and grow. Another major problem is the inefficiency of forming the desired specific cell type in the culture dish. While the process works quite well when the cells are left in an intact embryo, when taken out of that normal environment and placed into a lab culture dish, the embryonic stem cells do not behave as expected. They can be kept growing for long periods of time once started, but it is extremely difficult to get them to form just one particular desired cell type such as a neuron or heart cell. Instead the cells usually form a mixture of several different specialized cell types, along with some cells that just continue to grow. And it is these cells that continue to grow and divide that are potentially another big problem—if injected into a patient, they might form a **tumor**. This is in fact what happens in many of the mouse experiments; when injected into animals, the embryonic stem cells tend to form tumors along with a mixture of various differentiated cell types.

Thus far, there is also only scant evidence that embryonic stem cells can actually work in the body to successfully treat degenerative diseases. There are no current treatments available for human patients. Even in animal studies, the results are modest at best. Embryonic stem cells have been used in attempts to treat animal models of different human diseases, such as spinal cord injury, Parkinson's disease, and diabetes. The published scientific evidence shows some modest success at repairing spinal cord injury, allowing rats to regain some movement after treatment with embryonic stem cells. For Parkinson's disease, a little over half of rats treated with embryonic stem cells showed some improvement, but one-fifth of the rats treated died due to brain tumors caused by the embryonic stem cells injected into their brains. For diabetes, even though embryonic stem cells could be induced to secrete insulin in the lab dish, the cells didn't make enough insulin to prevent death of the diabetic mice that were treated.

Another potential problem for embryonic stem cell treatments is **transplant rejection**. Even with careful tissue matching, there is a possibility that implanted embryonic stem cells will be rejected by the body's immune system. To prevent the rejection of the transplanted cells and tissues, a patient would have to be treated with strong anti-rejection drugs for the rest of his or her lifetime, as is currently done for most normal organ transplants. Possibilities for minimizing the transplant rejection problem include genetic engineering of the implanted embryonic stem cells to match the patient's tissues, changing the patient's immune system itself by replacing it with a new immune system made from embryonic stem cells, or using embryonic stem cells made from an embryo that is a clone of the patient. Some of these techniques are only speculative, and so far none of these possibilities has been worked out successfully.

Another problem with using embryonic stem cells to treat diseases is that the cells themselves seem to be unstable in their ability to control their **gene expression**. The human genome contains anywhere from 30,000 to 60,000 different genes, which must all work together at the correct time and place during embryonic development so that the correct tissues and organs are formed in our developing bodies. Some published scientific research now shows that embryonic stem cells from mice show variability in how and when the different genes are turned on and off. The uncertain nature of how and when different genes are expressed may explain why it is difficult to get embryonic stem cells to form only one specific tissue, and why they tend to form tumors. This genetic variability will need to be carefully controlled if there is to be any potential for use of embryonic stem cells in clinical treatments.

ADULT STEM CELL ALTERNATIVES

While many still believe that embryonic stem cells are our best hope for treating various degenerative diseases, there are many others who point to alternatives such as adult stem cells as the more promising route for helping patients. As mentioned before, we have known about some of these adult stem cell types for years. Bone marrow transplants have been used routinely for treatments of various cancers and anemias. But until recently there was little evidence that adult stem cells had the same potential as embryonic stem cells. When the National Bioethics Advisory Commission studied the question of using human embryos to derive embryonic stem cells in 1999, their analysis of the science at that time led them to support the use of human embryos in research, but they also noted that this use had its limitations:

> In our judgment, the derivation of stem cells from embryos remaining following infertility treatments is justifiable only if no less morally problematic alternatives are available for advancing the research. . . . The claim that there are alternatives to using stem cells derived from embryos is not, at the present time [9/99], supported scientifically. **We recognize, however, that this is a matter that must be revisited continually as the demonstration of science advances.**
> "Ethical Issues in Human Stem Cell Research," National Bioethics Advisory Commission, September, 1999 [emphasis added]

So what does the science tell us now? Have any "less morally problematic alternatives" appeared? In short, yes, the published scientific literature is now full of reports on the abilities of adult and tissue stem cells with as much or more promise of treating disease as embryonic stem cells. Some of the past criticisms of adult stem cells have been that there was not a separate adult stem cell for each tissue, that they did not have the ability to make all body tissues, that they were difficult to isolate, and that they were present in the body in only small numbers and could not be kept growing indefinitely, so that it would be difficult to get enough of these cells to actually treat a patient. Since 1999, all of those ideas have been disproven in the scientific literature.

Adult Stem Cells Can Also Form All Body Tissues

In several studies, it's now been shown that adult bone marrow or brain stem cells can form virtually any body tissue (Figure 6 on page 12). In one experiment, scientists injected brain stem cells into an early mouse or chicken embryo (similar to what has been done with embryonic stem cells) and demonstrated that

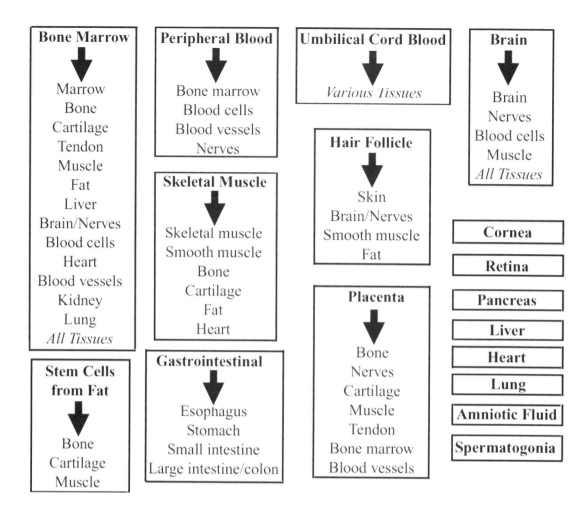

FIGURE 6. Some adult stem cells and tissues they can form.

the brain stem cell could help form many of the normal tissues of the animal. In another experiment, scientists destroyed the bone marrow of mice, then into these mice injected *one* bone marrow stem cell from another mouse. The recipients of this transplant would only survive if that one bone marrow stem cell survived and flourished (what might be called "mouse-to-mouse resuscitation"...) Not only did the single transplanted bone marrow stem cell survive and produce bone marrow and blood cells, but also could be found making liver, lung, digestive system, skin, heart, and muscle. And one research team has isolated what some have called the "ultimate stem cell"—a bone marrow stem cell that can grow forever in culture and make all body tissues. They did the "gold standard" test, injecting the bone marrow stem cell into an early mouse embryo, and finding evidence that the cell could help form all tissues of the body.

In the last few years a tremendous number of scientific reports have come out that show adult stem cells in most if not all body tissues. Basically, whenever someone has taken the time to look at a tissue, they've found an adult stem cell there. Hardly a week goes by without some new report of the isolation of another adult stem cell, or the ability of an adult stem cell to form another tissue type different than the one from which it was isolated. Figure 6 shows some of the examples. Bone marrow has been studied the longest and seems to be one of the best sources, being able to form all tissues of the body according to scientific reports, but there is obviously no shortage of possibilities. Even the brain has stem cells. This was a truly surprising finding. For at least 100 years, scientists believed that we started life with as many brain cells as we would ever have (and that it was downhill from there). But now we know that the brain contains stem cells that can be "awakened" to form more brain and nerve cells; some evidence indicates that this may be a continual process throughout our lifetime. And another surprise is that the brain stem cells are not limited to making just brain and nerve, but can also "cross-train" to form other tissues such as muscle and blood. Even fat has been found to contain adult stem cells that can form other tissues. So if we really need an unlimited source of stem cells, the United States is set!

How can scientists be sure in these experiments that it was the bone marrow stem cells that made the tissues, and not some other cell in the mouse's body? For the injection of the adult stem cell into the early embryo, specific genetic markers for the injected cell can be followed that distinguish the injected cell from the original embryo cells, in the same way that an injected embryonic stem cell was followed in the earlier experiments. For the adult stem cells transplanted directly into an adult mouse, the injected bone marrow stem cells not only have genetic markers that can be detected, but also the cells can be labeled with fluorescent markers that provide "glowing" evidence that the new tissue was formed from the adult stem cell (Figure 7 on page 14).

As far as isolation of adult stem cells, this would depend on the tissue from which you were getting the cells. So while it might be difficult to get out a brain stem cell so that you could make more nerve cells, it's relatively easy to get a stem cell from bone marrow, blood, or skin, all of which have been shown capable of making nerve cells.

Adult Stem Cells Can Multiply Almost Indefinitely

Even if there are very few adult stem cells available in the body to be isolated, the published scientific evidence now shows that adult stem cells can be kept

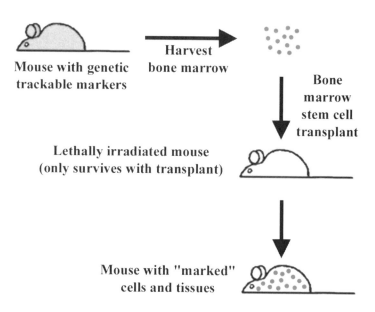

FIGURE 7. Detecting formation of tissues by marked adult stem cells.

growing almost indefinitely in culture. One research team was able to get a billion-fold increase in the number of adult stem cells in just a few weeks. Numerous other scientists have found key growth signals that can be added to cultures of adult stem cells to get them to multiply rapidly. And one group from the University of Minnesota has found that they can keep adult stem cells from bone marrow growing indefinitely. They found that not only did the cells continue to grow in their lab cultures, but that they kept going well beyond what had traditionally been thought of as the limits for growth of adult cells of any type. It had been thought for years that adult cells could only multiply a fixed number of times in culture, and then would **senesce** (undergo aging) and quit growing. Again, the new scientific evidence shows this is outdated thinking. The University of Minnesota team not only got their adult bone marrow stem cells to grow well beyond what was thought to be the limit for adult cell growth, but the stem cells didn't age at all and still retained the ability to form many different tissue types. So at least some adult stem cells do possess what are supposedly the two main advantages of embryonic stem cells—the ability to grow indefinitely so that many cells can be made for treatments and the ability to form any adult body tissue.

Adult Stem Cells Are Effective in Treating Animal Models of Disease

The real proof of the ability of any stem cell is not so much if it can keep growing or make different tissue in a lab dish, but whether it can actually repair damaged tissue, reversing the damage caused by disease. In this respect, adult stem cells have had good success, including with animal models of several of the diseases we would like to treat in people. Using bone marrow and muscle stem cells as well as umbilical cord blood stem cells, scientists have repaired damage in experimental animals due to heart attack, stroke, liver disease, diabetes, Parkinson's disease, and spinal cord injury.

Adult Stem Cells Are Successfully Treating Human Patients Now

If you remember, our main goal for stem cells was to use them to treat diseases in human patients, not just in animals. It's important that the treatments be tested first for safety and effectiveness, but we want to develop methods that will help patients suffering from diseases. In this respect, adult stem cells are already being used successfully to treat human beings for several diseases. Some of these treatments are a natural progression of many years using bone marrow transplants. In most cases, patients are now treated with bone marrow stem cells. This is an important treatment for many types of cancers, as well as for anemias, immune deficiencies, and other blood disorders. In addition, the same types of treatments are starting to be used for conditions including multiple sclerosis, lupus, rheumatoid arthritis, and scleroderma; these are called **autoimmune diseases**, because the patient's own immune system attacks the patient's body. In these treatments, some of the patient's bone marrow stem cells are removed and grown in the lab while the patient receives chemotherapy, which kills any cancer cells or malfunctioning immune cells but also destroys the patient's bone marrow. Then the stem cells are returned to the patient to regrow the bone marrow and restore the blood-forming system of the body. Many of these treatments are still experimental, but are showing great success at treating these diseases. In some cases, umbilical cord blood stem cells are used instead of the patient's bone marrow stem cells. Umbilical cord blood is an extremely rich source of blood-forming stem cells.

In some ways more exciting is the use of adult stem cells to form other tissues besides blood and bone marrow. These examples include forming new cartilage and bone, growing new corneas to restore sight to blind patients, treating stroke damage with neural stem cells, repairing heart muscle after a heart attack using the patient's own bone marrow or muscle stem cells (Figure 8 on page 16), and treating Parkinson's disease using the patient's own brain stem cells. In another example of the use of adult stem cells to treat patients now, a company in Switzerland will grow new skin for patients, starting with a few plucked

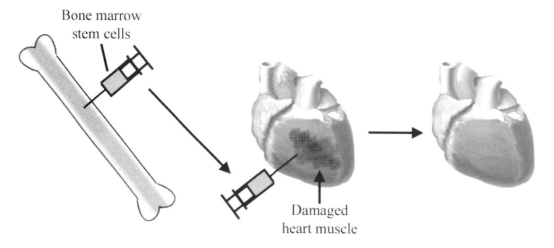

FIGURE 8. Using bone marrow stem cells to repair damaged heart muscle.

hairs from the patient! This is possible because the stem cell for skin is in the hair follicle. While the effectiveness of adult stem cell use is still being tested, so far the results are very encouraging.

NEW RESULTS, NEW IDEAS IN BIOLOGY

We need to keep in mind that most of these exciting developments in stem cell biology, especially with adult stem cells, have occurred just in the last few years. Because the results are so new, even many scientists are reluctant to accept them until they are more fully proven and repeated by several different scientists working in different laboratories.

This is an important part of the scientific method and "doing" science. While one laboratory may have great success in an experiment, before most scientists will accept that the results are true, the findings must be **replicated** in other, independent laboratories; that is, other scientists need to be able to repeat the experiments, in the same manner as originally done, and still get the same results to verify that those results are accurate. In a sense, it's a check against the possibility that one set of good results was simply a fortunate accident, instead of being a real mechanism by which the world works.

Another important aspect of science is publishing the results so that other scientists have an opportunity to learn, analyze the results, and add their own discoveries. The gold standard in publishing scientific results is that the information appears in a **peer-reviewed journal**. This means that the results have been reviewed by other scientists (peers) who have analyzed the results and agree that the experiments were done correctly, and that this is important new information that the rest of the scientific world should have an opportunity to see.

As scientists investigate the world, they try to develop explanations for how things (a cell, an organ, an animal, etc.) work and why they behave in certain ways. The scientist observes carefully, learns from other scientists and the scientific literature, and forms a **hypothesis**, a statement that attempts to explain how a particular thing works. Then experiments are designed to test the hypothesis to see whether it is really a true explanation. The results of the experiment help either **verify** or **falsify** the hypothesis (that is, is it a good explanation for the way something works, or not?). Again, the experiment must be repeated several times to make sure the answer is always the same. Over the course of many experiments, and other scientists replicating the experiments, the results mount either in favor of or against the hypothesis being true. If the evidence mounts that a hypothesis is true, it starts to become accepted as fact.

But this can actually work against science, certainly science in its best sense. Another aspect of science is always to question, to keep an open mind, to continue to test the hypothesis that's accepted as fact. It becomes almost tradition to believe that things are a certain way. Yet there is always another experiment to do, another test, even another hypothesis that may explain things better. New results may contradict the accepted thought. Hopefully this leads to changing ideas, and a new **paradigm**, or way of looking at things.

Stem Cells and Development—Old and New Ideas

How we develop from a single-celled embryo into an organism of 100 trillion cells, all working together but each with specialized tasks, has been a fascinating area of science and continues to be largely a mystery. One of the original hypotheses was that a tiny preformed human being was present in the egg or the sperm, and that when the two were joined this tiny human being just started growing larger and larger. Of course, then microscopes got better and no one could see the tiny person in the cell!

When an embryo develops from that one-cell **zygote** (the one-celled embryo) on to the blastocyst stage, some cells have already taken on certain specific jobs. The outer layer (trophoblast) goes on to form placenta, and the inner cell mass goes on to form all body tissues. At the next stage of development, those inner cells separate into three layers (the three primary germ layers— ectoderm, mesoderm, and endoderm), which each form different body tissues, as discussed before. From there, cells become more and more specialized in their job descriptions until the final differentiated tissues are formed. It's a bit like going up a tree, taking one of three main branches from the trunk, and then going farther and farther out on the limbs (Figure 9).

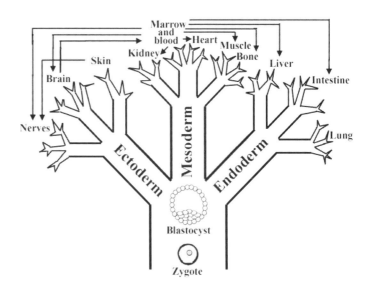

Marrow
and
blood →Heart
Kidney Muscle
Skin Bone
Liver
Brain
Intestine
Nerves
Lung
Ectoderm Mesoderm Endoderm
Blastocyst
Zygote

FIGURE 9. Old and new ideas about how cells develop.

The traditional idea in developmental biology has been that, as a cell, the farther you went out on a limb, the more "stuck" you were, that you couldn't back down the branch and take a different branch to become a different type of cell or tissue. It's a bit like the proverbial cat stuck up a tree, which can't or won't come down and necessitates calling the fire department to get Fluffy down. Going from one of the three main branches to a different main branch was considered simply impossible. Then along came adult stem cells, which seemed to have the ability to move among different branches. Some people started thinking that maybe the cells were backing down the branches after all, then climbing back up.

But current scientific evidence indicates something else completely. Instead, adult stem cells seem to be behaving not like cats, but instead like squirrels jumping from branch to branch. This is a totally new idea, a new paradigm or way of looking at things, that doesn't fit with what scientists had previously thought. The adult stem cells seem to look at what tissue they're in and simply become that tissue type. One scientist, Helen Blau from Stanford University School of Medicine, has proposed that there may be only *one* type of adult stem cell, found in many or all tissues. This adult stem cell can leave one tissue and wander in the bloodstream (she calls the circulatory system "route stem cell"). Then the stem cell gets out in a different tissue, something like taking the off-

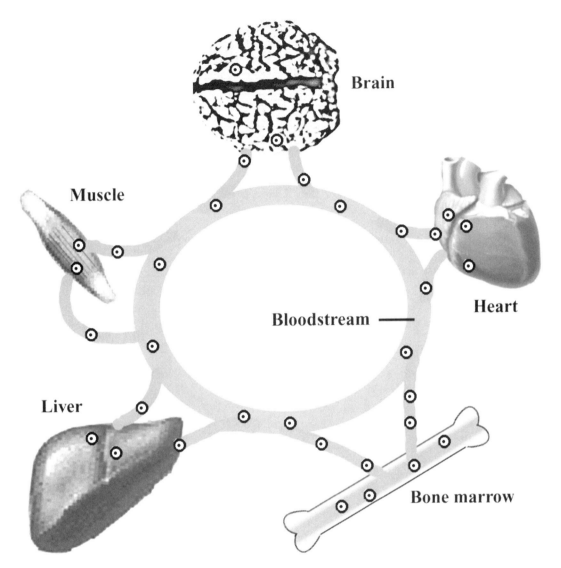

FIGURE 10. The stem cell superhighway with cells moving among organs.

ramp at a freeway exit (Figure 10). It reads the signs as it gets off and gets into the tissue, sees what the neighborhood is like, recognizes the signals that this is a different tissue, and becomes that type of tissue. If it goes into the brain it becomes a nerve cell, into the heart it becomes a cardiac cell, and so on. There is actually very good evidence now that adult stem cells do migrate through the bloodstream. However, there are still many scientists who are skeptical. Anytime traditional ideas are challenged, the new ideas are slow to be accepted.

One other argument for using embryonic stem cells is that they may be useful to figure out how development proceeds, and how adult stem cells are formed. There are both advantages and disadvantages to this proposal. Embryonic stem cells could be helpful in determining the sequence of events that leads to specific tissues or adult stem cells. This is an interesting basic science question. However, if the goal is treating patients, adult stem cells have already been isolated and grow well in culture, so they could be studied directly to discover the signals needed to turn them into different tissues needed to treat diseases, without the need to go through previous steps of development.

Cloning

When people talk about cloning, most mean reproducing an entire organism. Many think about cloning a person from the past, such as Elvis, or envision science fiction movies in which armies of cloned soldiers take over the world. The real story is far different. In fact, the word cloning simply means to make an identical or nearly identical copy. Molecules such as DNA can be cloned or individual cells can be cloned—these cloning processes make multiple copies of the

FIGURE 11. Dolly, the first mammal cloned from an adult cell nucleus.

original DNA molecule or cell. But when an organism is cloned, the complete adult animal is not immediately produced. The clone starts as a one-celled embryo, and must still go through an embryonic stage, grow, and develop. Frogs were first cloned in the 1960s, but it was not until 1996 that the first mammal, Dolly the sheep (Figure 11), was cloned using an adult cell nucleus. Dolly was born in 1997.

THE HOW AND WHY OF CLONING

Cloning an organism does not immediately give you a full-grown adult. Dolly the sheep started as an embryo, just as other sheep. The difference for Dolly and other animal clones is that they were not produced by **sexual reproduction**—that is, the union of egg and sperm—so the process is termed **asexual reproduction**. In normal sexual reproduction, an egg and sperm are joined together. Each brings half of the chromosomes for the new individual, so that the one-celled embryo contains a full set of chromosomes (half from mom and half from dad; Figure 12). The process of cloning an organism requires two main ingredients: (1) an egg cell, and (2) the genetic material from a body cell (**somatic cell**) of the organism to be cloned. The chromosomes (genetic material) are first removed from the egg, a process called **enucleation** (or the genetic material can be inactivated by radiation). Then the nucleus containing the

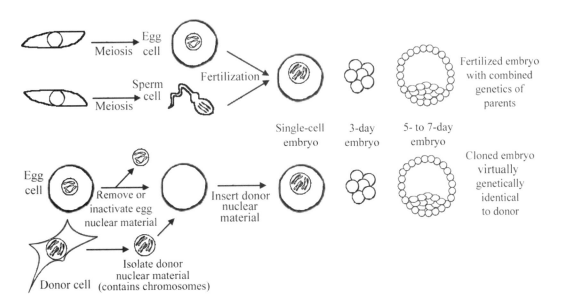

FIGURE 12. Making embryos—sexual fertilization versus cloning

genetic material is taken from a body cell and inserted into the enucleated egg. In some cases, rather than removing the nucleus from the body cell, the whole cell is fused together with the egg to make a single cell.

This cloning process creates a one-celled embryo, but in this case all of the chromosomes come from the cell that supplied the nucleus put into the empty egg. Other terms for this process are **nuclear transplantation** or **somatic cell nuclear transfer**. The new clone is then stimulated to grow and develop, sometimes by applying an electric shock and sometimes with chemicals. The embryo can grow for a few days in the laboratory dish, but for it to continue development the embryo must be implanted into the womb of a **surrogate mother** and undergo normal gestation to birth. An embryo produced by sexual reproduction has the combined genetic makeup of the two parents. An embryo produced by cloning is virtually genetically identical to the donor of the genetic material.

Purposes of Cloning—Why Do We Want to Clone?

So why would we want to make a clone? For agricultural uses there are many reasons why having cloned animals could be an advantage. Selectively breeding the best livestock—for example, crossing the best bull with the best cow—can produce champion animals, such as good milk or wool producers. But when that champion animal is bred there is no guarantee that its offspring will have all of the same characteristics, because only half of the genetic makeup of the offspring will be from that champion animal. However, if we could produce clones of that animal, then all of the cloned offspring would have the same genetic makeup of the champion. It is even possible to genetically engineer animals so that their milk contains medically valuable human proteins such as insulin or clotting factor, or perhaps so that their organs could be used for transplantation and not be rejected by human organ transplant patients. For these types of animals, cloning could ensure that there are many identical animals available. Cloning could also be useful in producing **animal models of disease**—animals that have a specific disease—so that the medical condition can be studied in attempts to find treatments and cures.

For human cloning there are two purposes proposed. One has been termed **reproductive cloning** or **live-birth cloning**. Here the idea is that the cloned human embryo would be implanted into a womb and develop to be born. Proponents of this purpose for human cloning talk about using it to help infertile couples to have a child, or to reproduce a child who has died. The other purpose for human cloning has been called **therapeutic cloning** or **experimental cloning**. The idea here is to clone an early embryo of a patient who has a degenerative disease, then use the embryo for production of embryonic stem cells, in hopes of treating the patient. In theory, the cells will then be a genetic match for the patient and will not cause transplant rejection. Some people also call this cloning technique **nuclear transfer to produce stem cells**. However, the process is still cloning, and the product is still a human embryo. The difference between the two reasons for human cloning is simply the purpose for which the embryo is used—a born child or embryonic stem cell treatments.

Reproductive Cloning

Reproductive cloning—cloning to produce a live birth—has been accomplished with several animals thus far. Since Dolly was cloned, other animals that have also been cloned and brought to birth are mice, goats, pigs, cattle, rabbits, and a cat ("cc," the carbon copy cat). There have been attempts with other animals, including dogs, monkeys, wild cats, and a rare species of wild ox, but these clones either did not survive to birth or died not long after birth. This points out one of the big problems in cloning—most clones don't survive. Frankly, it's unsafe to be a clone!

Cloning, as it turns out, is a very inefficient process. It took 277 tries to get one Dolly. Each attempt involved creating a cloned embryo by the nuclear transfer technique. Mammary gland cells from a 6-year-old sheep were used as the nucleus donor, fusing these cells with an enucleated egg cell to create the one-cell clones. Only about 10% of these one-cell cloned embryos develop to the blastocyst stage (in Dolly's case, only 29 of the 277 cloned embryos). Those that reach the blastocyst stage can be implanted into a **surrogate mother** for gestation. Most of these implanted embryos don't develop or survive to birth; of the 29 implanted embryos, only Dolly was born.

The numbers are not much better for other cloned animals. For mice, one experiment created 613 cloned embryos to get five born mice, and in another experiment it took almost 1,000 cloned embryos to get two born mice. With rabbits, 1,852 cloned embryos were made but only six rabbits were born. Bigger animals seem to do better, but not by much. It took 72 cloned embryos to get five born pigs, 85 cloned embryos for three born goats, 496 embryos to get 24 born cattle, and 188 clones to get one live kitten. And these are the survivors who made it to birth and lived to grow up. Many don't make it to birth. For the cloned cat, only 87 of the 188 cloned embryos developed to the blastocyst stage, where they could then be implanted into a womb. Others die before or soon after birth, and seem to have various abnormalities with their organs.

There is also a question as to whether any clones are normal, even those that survive and grow to adulthood. Some experiments show that Dolly is actually older than her birth age; she was cloned from a 6-year-old sheep, and may be older genetically than she should be. One indication of advanced age is that she has developed early onset arthritis, earlier than a normal sheep should. Another laboratory found that most of their cloned mice died early.

While many surviving clones have gone on to reproduce normally, including Dolly, some experiments indicate there may be some fertility problems with cloned animals.

Another aspect of bringing a clone to live birth is the surrogate mother who carries the clone. Remember that the cloned embryo has to be implanted into a womb to continue development to birth. Virtually all clones are born by Caesarian section, to protect the health of the female carrying the clone. This is because the clones, and their placenta, usually grow faster and larger than normal, termed **large offspring syndrome**. It's not known exactly why this overgrowth occurs, but it could have something to do with the way different genes are expressed during embryological development.

Normal development is an intricate dance of genes being expressed at the right time and in the right cells. The problem with getting normal clones probably has to do with **gene reprogramming** when the nucleus is transferred into the egg. In sexual reproduction, the egg and sperm each have their genes programmed to start normal development. The way that genes are set to be expressed is called **gene imprinting**. Think of the genes as a set of on/off switches; different switches will be set off or on, depending on the particular cell type, tissue, or stage of development. When the nucleus of an adult body cell is transferred into the enucleated egg to create a clone, many of the switches are set differently than they would be for the start of embryo development, because the body cell has its programming set to do its job in a particular tissue. The cytoplasm in the egg must reprogram the genes, resetting the switches so that normal development can occur. The failure of most cloning attempts likely is a result of not getting all of the switches set correctly. One laboratory that studied several genes during mouse cloning found that this was the case—each of the mice born showed variability in expression of the few genes that were examined. In this particular case the differences were not enough to prevent a few of the mice from developing to birth, but with thousands of genes this can be a serious problem. It is also uncertain what such changes in the gene programming might mean later in the life of the clone.

Most scientists and the general public oppose attempts to clone a human for a live birth. The National Academy of Sciences has recommended that this should not be done at present, simply because it is unsafe.

Therapeutic Cloning

Therapeutic cloning—cloning to produce embryonic stem cells for medical therapies—is probably an even more controversial proposal. The idea is to provide therapies for patients with diseases such as Parkinson's disease, diabetes, and so on, using embryonic stem cells. By producing a cloned embryo of the

clinics. For **in vitro fertilization (IVF)**, a woman is given high doses of hormones to mature many eggs at once—ten, twenty, even thirty. These eggs are then surgically harvested and fertilized in a lab dish. The purpose of IVF is to help a couple conceive a baby. After fertilization and growth in a dish for a few days, anywhere from one to six of the embryos are implanted in the woman's uterus, and the rest of the embryos are frozen. These frozen embryos are essentially a hedge on the bet that one of the implanted embryos will develop and be born. If that does not occur, some of the embryos are thawed and the couple tries again. Or after the initial birth, they may want more children. Nevertheless, there are inevitably some embryos that are not used and are left in the freezer. They can survive for extremely long periods (the evidence so far is that they don't seem to go bad or get freezer burn like a steak). The estimates are that somewhere around 100,000 human embryos are in freezers in the United States alone.

Inevitably some of the frozen embryos are discarded. Fertility clinics obtain a signed consent form from the couple, and one option allows the clinic to discard embryos after several years if they are not used. However, most are not discarded. What to do with all the frozen embryos? They may be donated for research. This was the bone of contention in the stem cell debate. Should human embryos be used for research, given their original purpose was to be implanted for a birth? Why not use them for research if they would be discarded (with the parent's permission) anyway?

Another option for frozen embryos is relatively new—embryo adoption. The **Snowflakes Embryo Adoption Program** sets up adoptions for embryos whose genetic parents have achieved their family. This option, while relatively new, is growing rapidly and offers a lifesaving option for frozen embryos. Parents of adopted embryos have testified before Congress regarding this option and against embryo destruction.

While accumulating so many embryos in freezers is a problem, newer techniques often don't create as many embryos as in the past. One solution is only to create as many embryos at one time as will actually be implanted. Another solution is to freeze eggs, which can then be thawed, fertilized, and the embryos implanted as needed.

Creating Embryos from Scratch

Human cloning raises even more questions about what it means to be human. Cloning involves the specific creation of human embryos with certain ends in mind. For "reproductive cloning" there are worries beyond the safety factors. For a couple who create a cloned child, a clone of the wife will not be the genetic daughter but instead will be the sister, a late-born twin, and will not be related to the husband. How will this change kinship and family relationships? What will be the expectations put on a clone once born? And what if—which is possible—a previously existing person, now deceased, is cloned? The genetic makeup of a clone will already be known, already dictated because the process of cloning reproduces a previously existing individual. Will the clone be expected to live up to that genetic legacy? Will there be heightened expectations by the parents and others based on what was achieved by the donor of the genetic material that made the clone?

However, keep in mind that even though the genetic makeup of a clone is predetermined, there are many other factors that go into our overall composition. Our genes determine many of our physical characteristics and even predispose us to various diseases or behaviors, but we are also products of our environment and experience. A good example is the cloned cat "cc." Even though she looks very similar to the cat who donated her genetic material, her coat pattern is slightly different. This is because even the environment in the womb can affect development—in this case, coat pattern. Identical twins have different fingerprints. And of course after we are born there are many experiences and environments that make us who we are and who we will become. Those experiences can't be duplicated, so the clone will grow up differently than the one who was cloned, and may behave quite differently. A clone of Einstein might become an artist instead of a scientist. We are so much more than just our genes!

Creating cloned embryos for experimental or medical use raises the questions posed before regarding moral status. Should humans (realized or potential) be created and destroyed for the potential benefit of others? Again there are a range of viewpoints regarding the status of the embryo. If the embryo at this early stage does not have as high a value, or any value, compared to other human life, then **utilitarian logic** (the idea that something is only good if it is useful, and that actions should promote the greatest good for the greatest number) would dictate that it should by all means be used. Others argue that it should be protected, not because of its inherent value but because creation of human embryos for such purposes can lead to human commercialization, making any human life a commodity to be bought, sold, and used, cheapening life. Still others would say that we should not create human embryos for purposes other than reproduction, and not in a manner that manufactures human life, so-called "designer embryos." The question still goes back to what it means to be human and what value is placed on human life.

Medical Breakthroughs

If the goal of medical research is to provide treatments for patients with diseases, then it would make sense to look at how well different types of stem cells can provide those treatments. Theoretically, embryonic stem cells could provide a one-size-fits-all solution, able to be grown on demand into any needed tissue to repair damage from disease or injury. But can they actually produce these tissues in the laboratory? Can adult stem cells potentially do everything that embryonic stem cells might do? If adult stem cells and other alternatives can provide all of the clinical treatments for patients, do we need embryonic stem cells at all? This is a key part of the debate.

Scientific Freedom, Scientific Stewardship

Part of the debate also focuses on science, but from the viewpoint of science's role in society. Science and technology have provided us with wonderful inventions and discoveries that make our lives happier and healthier. Should science have unlimited freedom? It's often difficult to know from where the next major breakthrough may come, and sometimes new discoveries arise from unexpected sources and paths that even many scientists may not consider worth following. Should scientists be allowed to do whatever they feel interesting, or whatever they feel may provide the most benefit to society? Or should science follow the lead of what society feels is best, or what policymakers believe is best? How should science best serve society? The whole concept of regulating something as unpredictable and free flowing as scientific discovery is a difficult one. It is also difficult to know who should decide these questions—the scientists because they know how science works, the policymakers because they must set the rules, or society because they are most affected by the decisions.

Economics

Like it or not, money plays a big role in these decisions. Private companies invest huge sums in hopes of a profit. Patents are filed, investors lined up, and there is money to be made in biotechnology and potential medical treatments. Science is not immune to these pressures. Not only is scientific research expensive, but projects and whole careers may live or die depending on public grant funding and private investment in research. Federal funding plays a big role simply because the government invests a great deal of money in scientific research and health care. This was actually at the heart of the original debate that led to President Bush's August 9, 2001 decision regarding embryonic stem cell research. The question was not about the legality of using human embryos for research, but only whether the Federal government would fund such research. And even though the amounts of money seem tremendous, there is still only a limited budget for all of the possible research. Should scientific research be funded based on its likelihood of producing disease cures or other useable knowledge? An additional aspect of economics is how much treatments will cost and how available such treatments might be for the average person. If any treatments developed will be so expensive that only the rich can afford them, should other avenues be explored instead?

Politics

In such contentious and divisive issues, with prestige and money at stake, politics and policymaking are in play as well. The issues of stem cells and cloning have generated a great deal of debate, Congressional hearings, speeches and rhetoric. But the ongoing debate is helping everyone learn the issues and make their voices heard.

WHAT ARE OTHER COUNTRIES DOING?

Not surprisingly, stem cells and cloning are issues for debate not just in the United States but around the world. Much is at stake in terms of possibilities for scientific advance, economic and biotechnology development, and national prestige. As you might expect, the decisions span the range of potential outcomes, though for the most part the debates are continuing without final outcomes yet in many countries. Great Britain has one of the most liberal policies regarding stem cell and human cloning research. They allow embryos to be created, whether by fertilization or cloning, and used for research up to 14 days after conception. This would include "therapeutic cloning" research, though their law prohibits implantation into a woman, thus banning reproductive cloning. France falls somewhere in the middle, currently allowing use of "excess" embryos from fertility clinics for research, but not the creation of embryos for research. Germany has some of the most restrictive laws, forbidding creation of embryos for research, including by cloning, and also prohibiting destruction of "excess" embryos for research purposes. German researchers can import human embryonic stem cells if there are no alternative ways to conduct the research.

Our next-door neighbor, Canada, has been going through the same debates as the United States. A legislative proposal has been made to allow use of frozen embryos for research, but not creation of embryos by fertilization or cloning for any research purposes. A few countries have banned all human cloning, but virtually all countries are still debating the use of human embryos. In the United Nations, there are proposals for a global ban only on reproductive cloning, or on all human cloning.

THE DEBATE IN THE UNITED STATES

In the United States, the main focus has been on Federal funding. The Clinton administration had approved guidelines for funding of research with human embryonic stem cells, but this was blocked by a lawsuit. President Bush decided on August 9, 2001, that government funds could be used for research on human embryonic stem lines that were already in existence (over 70 such cell lines are believed to meet the conditions) on that date, but that no funding would be given for embryo destruction, effectively preventing funding of any new cell lines developed after that date. At this time, it is legal in most states to use human embryos for research purposes (as long as there are no Federal funds involved), and legal in most states to perform human cloning.

The Federal government does not allow funding for the creation or destruction of human embryos for research. This prohibition has been in place since 1996, as a part of the annual funding bill for the Department of Health and Human Services, and is known as the "Dickey-Wicker Amendment" (after the members of Congress who proposed the amendment). The language of the amendment states that "the term 'human embryo or embryos' includes any organism . . . that is derived by fertilization, parthenogenesis, cloning, or any other means from one or more human gametes or human diploid cells."

Various legislative bills are under consideration in the U.S. Congress regarding stem cells and cloning. These include bills that would authorize funding of embryo research as well as bills that would prohibit all embryo research and increase funding for adult stem cell research. Competing bills are also under consideration regarding human cloning. The House of Representatives passed the Weldon-Stupak Bill in the summer of 2001 by over a hundred vote margin. The bill would prohibit all human cloning in the United States. The Senate version of that bill, the Brownback-Landrieu Bill, is under debate in the spring of 2002, along with the competing Specter-Feinstein-Kennedy and Dorgan Bills, which would prohibit only reproductive cloning. President Bush has endorsed a ban on all human cloning.

Michigan, Virginia, and Iowa have banned all human cloning. Several states (9 to 10) currently have laws against embryo research, some of which would prevent therapeutic cloning since embryos are destroyed. Almost all states are debating the issue and have bills under consideration regarding cloning and embryo research. The individual states may be the ones that move faster in terms of the legislation, which may affect the debate on the Federal level.

Everyone into the Debate!

Virtually everyone has heard of stem cells and cloning, and many people have formed an opinion on these types of research. Scientists obviously have a stake in the debate, in terms of what research they will or won't be allowed to pursue, who might fund the research, and possibly where they can do the research. Patient groups also have a big stake in the outcome of these debates. They're interested in treatments and cures! And they want to know what type of research will actually benefit them most. Various moral and religious leaders and groups are also heavily involved in the discussion, because at one of the most basic levels the debate centers around a moral or ethical question. Politicians obviously play a big role in the debate, because they propose and vote on the legislation and fund bills that will determine the outcomes. Another group involved that we don't usually think about is the media. Their role is to report the straight facts, though sometimes they may not have all of the facts or may actually take sides.

An interesting byproduct of these biology debates are the new coalitions that are forming. For example, you might think that the question of using embryos for research is an extension of the abortion debate, with pro-life and pro-choice forces lined up against each other. However, the debate regarding the early embryo is not the abortion debate. A woman's choice of whether to allow her body to gestate the embryo is not in question. The debate centers around an embryo in a petri dish or freezer and the question of creating embryos for various purposes. In the cloning debate, an interesting left-middle-right coalition has formed from pro-choice, pro-life, environmental, and other groups who all support a total ban on human cloning. The debates are new and so are the debating groups.

The Future

What's in the future for these issues? There is a lot yet to come. Even if some of the legislation is approved and decisions are made now regarding stem cells, embryos, and cloning, most of these issues won't go away anytime soon. There will be more debates, more legislation, legal challenges, and especially more science to discuss. Scientific evidence will accumulate one way or the other for the current debates, but the basic questions will continue to challenge us.

More questions are coming, because more scientific discoveries and potential applications are on the horizon. Some of these include additional ways to make embryos, which will keep bringing us back to questions of what makes an embryo an embryo. Another interesting technology that will spur debate involves **chimeras**, hybrids of humans and animals. These might be hybrids in which animal cells are placed into humans, or hybrid animals that contain large numbers of human cells, or hybrids in which human and animal genes are combined. We already have mutant mice with human genes, created so that we can study specific types of human diseases. But there is also the possibility of adding animal genes to humans, perhaps to help fight a disease or provide other capabilities.

Genetic engineering is also on the horizon. The first successful treatments of individuals with genetic diseases have been accomplished, replacing a defective gene with a working normal gene. This was done by **somatic genetic engineering**, in which somatic cells (body cells) of an individual are treated. But there is also the possibility of **germline genetic engineering**, in which the egg, sperm, or early embryo is genetically altered. This could be done to remove a disease gene or replace a defective gene, but the same techniques can be used to enhance the genetics of an individual. And with germline genetic engineering, the new genetic information not only is present in the individual who is treated, but also is passed down to all future generations. Finally, another advance on the horizon is **cybergenetics**, combining genetic advances with technological and electronic advances to heal or even enhance human beings.

It's truly an interesting and challenging future, and it's time you got into the debate, too! The decisions made now can literally affect the future of the human race. Hopefully this booklet on the basics of stem cell research and cloning has given you a start.

The future of the science of regenerative medicine is very exciting, and as a scientist and professor I hope you've enjoyed learning more about this important area of biology. I would welcome hearing from you with your thoughts about this booklet, and about stem cells and cloning. Please feel free to contact me at the following address. Best wishes for your studies.

David Prentice
Department of Life Sciences
Indiana State University
Terre Haute, IN 47809
E-mail: prentice@indstate.edu

Resources

Because the information on these issues is relatively new and changing rapidly, websites provide some of the best up-to-the-minute information regarding background as well as current knowledge. The following are some of the best across a wide spectrum of viewpoints.

Americans to Ban Cloning A group of concerned Americans and U.S.-based organizations that promote a global ban on human cloning. Website contains various information related to human cloning. — http://www.cloninginformation.org/

Bioethics.net The American Journal of Bioethics Online; lots of articles and basic discussion of bioethics topics. — http://bioethics.net/

The Center for Bioethics and Human Dignity The Center exists to help individuals and organizations address the pressing bioethical challenges of our day, including managed care, end-of-life treatment, genetic intervention, euthanasia and suicide, and reproductive technologies. — http://www.cbhd.org/

Council for Responsible Genetics A non-profit/non-governmental organization devoted to fostering public debate about the social, ethical, and environmental implications of the new genetic technologies. — http://www.gene-watch.org/

Do No Harm: The Coalition of Americans for Research Ethics A national coalition of researchers, health care professionals, bioethicists, legal professionals, and others dedicated to the promotion of scientific research and health care that does no harm to human life. Extensive website with information related to stem cells. — http://www.stemcellresearch.org/

The Hastings Center for Bioethics The Hastings Center is an independent, nonpartisan, interdisciplinary research institute that addresses fundamental ethical issues in the areas of health, medicine, and the environment as they affect individuals, communities, and societies. — http://www.thehastingscenter.org/

National Bioethics Advisory Commission Former advisory commission prior to the current President's Council on Bioethics. Website contains reports on stem cell research and cloning. http://bioethics.georgetown.edu/nbac/

National Institutes of Health Bioethics Resources on the Web Bioethics resources, links, tutorials. http://www.nih.gov/sigs/bioethics/casestudies.html

National Institutes of Health Report—"Stem Cells: Scientific Progress and Future Research Directions" Review of the science and challenges of stem cells; June 2001. http://www.nih.gov/news/stemcell/scireport.htm

National Institutes of Health Stem Cell Information Basic information on stem cells, stem cell primer, research guidelines, approved stem cell lines, and other information. http://www.nih.gov/news/stemcell/index.htm

National Reference Center for Bioethics Literature Information, references, and links to bioethics literature. — http://www.georgetown.edu/research/nrcbl/

National Research Council Report— "Scientific and Medical Aspects of Human Reproductive Cloning" Extensive overview of science involved in human cloning; January 2002. — http://www.nap.edu/catalog/10285.html

National Research Council Report—"Stem Cells and the Future of Regenerative Medicine" Extensive overview of the science and issues of stem cells; September 2001. http://www.nap.edu/catalog/10195.html

New Scientist Special Report on Stem Cells and Cloning Cloning and stem cell information and news reports.— http://www.newscientist.com/hottopics/cloning/

PhRMA Genomics News, legislative news, general information from a bio-pharmaceutical viewpoint; links to cloning, stem cell, bioethics, and other health information. — http://genomics.phrma.org/

President Bush Remarks on Human Cloning Legislation Transcript of April 10, 2002 speech. http://www.whitehouse.gov/news/releases/2002/04/20020410-4.html

President Bush Remarks on Stem Cell Research transcript of August 9, 2001 speech. http://www.whitehouse.gov/news/releases/2001/08/20010809-2.html

The President's Council on Bioethics Formed November 28, 2001. The Council advises the President on bioethical issues that may emerge as a consequence of advances in biomedical science and technology. — http://bioethics.gov/

Stem Cell Research News Independent reporting organization not affiliated with any advocacy, trade, research, or governmental organization, dedicated to providing independent, authoritative, current, objective reporting on all facets of stem cell research, including human embryo and adult stem cell research, to all interested parties. http://www.stemcellresearchnews.com/

University of Pennsylvania Center for Bioethics The Center's mission is to advance scholarly and public understanding of ethical, legal, social, and public policy issues in health care. http://www.med.upenn.edu/bioethic/

Yahoo News Coverage—Cloning Provides links to news, editorials related to cloning.—http://dailynews.yahoo.com/full_coverage/tech/cloning/

Yahoo News Coverage - Human Stem Cell Research Provides links to news, editorials related to ste cell research. http://dailynews.yahoo.com/full_coverage/science/stem_cell_research/

Your Genes, Your Choices: Exploring the Issues Raised by Genetic Research
Department of Energy site provides thought-provoking scenarios on ethical, legal, and social issues created by genetic research.
http://www.ornl.gov/hgmis/publicat/genechoice/index.html

FURTHER SUGGESTED READING

"Biological Alchemy" by W. Wayt Gibbs. *Scientific American.*
http://www.sciam.com/2001/0201issue/0201scicit1.html
"Mother Nature's Menders" by Mike May. *Scientific American.*
http://www.sciam.com/specialissues/0600aging/0600may.html
"Our Posthuman Future: Consequences of the Biotechnology Revolution"
by Francis Fukuyama. (Farrar, Straus & Giroux, 2002.)
"Redesigning Humans: Our Inevitable Genetic Future" by Gregory Stock.
(Houghton Mifflin Co., 2002.)
"What Clones?" by Gary Stix. *Scientific American.*
http://www.sciam.com/explorations/2001/122401clone/index.html